STARS

WHIMSY,

WISDOM,

AND LIGHT

FROM

THE OTHER SIDE

OF THE

DAY.

NORM KOHN

PEACHTREE PUBLISHERS, LTD. / ATLANTA, GEORGIA

PUBLISHED BY
PEACHTREE PUBLISHERS, LTD.
494 ARMOUR CIRCLE, NE
ATLANTA, GEORGIA 30324

TEXT AND ILLUSTRATION © 1992 BY NORM KOHN

MANUFACTURED IN THE UNITED STATES OF AMERICA

1 0 9 8 7 6 5 4 3 2 1

LIBRARY OF CONGRESS CATALOGING-IN PUBLICATION DATA

Kohn, Norm, 1935-
    Stars: Whimsy, Wisdom, and Light from the Other Side of the Day /
Norm Kohn.
        p. cm.
    ISBN 1-56145-064-2: $14.00
    1. Kohn, Norm, 1935- . 2. Stars in art. 3. Meditations.
I. Title.
NC975. 7. K64A4 1992   92-25247
741.6 '4' 092 --dc20   CIP

FOR
KATHY & ALAN
AND
BRYAN, GRACE & ROBERT

STARKEEPERS

STARS
MAY
RISE
ANYWHERE,
ANYTIME,

BUT NEVER, EVER IF YOU ARE WATCHING FOR THEM.

KOHN

STAR

LIGHTERS

ARE SPECIAL PEOPLE. YOU WILL KNOW THEM BY THE WAY THEY TURN ON THE

NIGHT.

STAR

    CATCHING

        CAN

         BE

          A

           WONDERFUL

            GAME.

THEY WILL ALWAYS LET YOU THINK THAT YOU CAN

            KEEP

            THEM.

THE NEXT TIME YOU FEEL UNNOTICED, TAKE YOUR STAR FOR A WALK.

THE
PROBLEM
WITH
TAKING
YOUR
STARS
FOR
A

SUNDAY EVENING DRIVE IS THAT THEY ALWAYS WANT TO TAKE THEIR FRIENDS ALONG.

ONE

OF

LIFE'S

MOST

REFRESHING MOMENTS   IS GETTING CAUGHT OUT IN A STAR SHOWER.

WHEN THINGS GET SO BAD THAT YOU KNOW YOU'LL NEVER MAKE IT, SNEAK UP ON A STA

AND

HUG

IT.

YOU'LL
KNOW
THAT
THE
FIRST
FROST
IS JUST
AROUND
THE
CORNER
WHEN
YOU
SEE
STARS
FLYING
SOUTH
FOR
THE
WINTER.

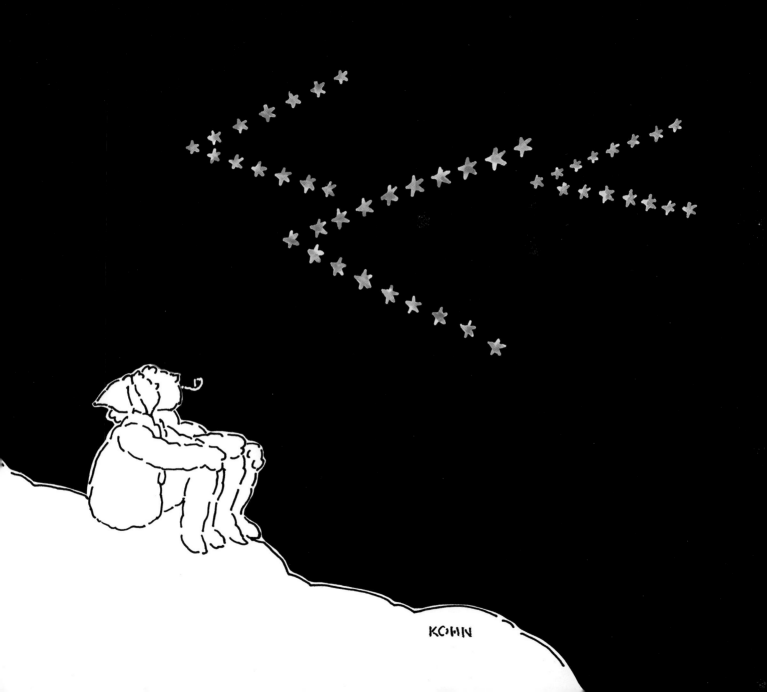

ISN'T IT AMAZING HOW F   R   E   S   H   LIFE APPEARS WHEN

YOU HAVE ON YOUR STAR - COLORED GLASSES?

THERE'S

NOTHING MORE FUN

ON

A

BLUSTERY

MARCH

EVENING

THAN TO BE OUT ON YOUR FAVORITE HILL FLYING A STAR.

WHEN YOUR HEARTS SAY IT'S TIME TO SLIP AWAY, GO FOR A PICNIC IN A FIELD OF STARS

A

STAR-TAN

IS JUST THE TICKET FOR THAT AFTER-SUNDOWN

G   L   O   W.

LEAVING A FAVORITE PLACE ISN'T SO DREARY IF YOU LET YOUR STAR PLAN THE TRIP

YOU'LL KNOW

YOU'VE FOUND WHAT YOU HAVE

ALWAYS BEEN LOOKING FOR

WHEN

YOU LOOK IN THE MIRROR

AND SEE A

STAR.

ABOUT THE AUTHOR:

NORM KOHN IS AN AWARD-WINNING DESIGNER
AND ILLUSTRATOR, WORKING IN PRINT, FILM AND
TELEVISION. HE HAS TAUGHT AT AUBURN UNIVERSITY,
THE ATLANTA COLLEGE OF ART, GEORGIA STATE
UNIVERSITY AND THE PORTFOLIO CENTER. HE
ILLUSTRATED "NIGHT WATCH" BY PAUL DARCY BOLES
AND WROTE AND ILLUSTRATED "BEGINNINGS,"
WHICH WAS CHOSEN ONE OF THE FIFTY BOOKS OF
THE YEAR FOR 1967 BY THE AMERICAN INSTITUTE
OF GRAPHIC ARTS. A RESIDENT OF ATLANTA, HE
ENJOYS YOGA, SAILING, AND CHAMOMILE TEA. "STARS"
EMERGED FROM HIS ONGOING INTEREST IN THE CREATIVE
RESULTS OF NONCAUSAL THINKING.